YOUR KNOWLEDGE HAS VALUE

- We will publish your bachelor's and
  master's thesis, essays and papers

- Your own eBook and book -
  sold worldwide in all relevant shops

- Earn money with each sale

Upload your text at www.GRIN.com
and publish for free

Lisa Nathalie

A study on Aspergillus flavus

Biochemical characterization of Aspergillus flavus

GRIN Verlag

Bibliografische Information der Deutschen Nationalbibliothek:

Die Deutsche Bibliothek verzeichnet diese Publikation in der Deutschen National-
bibliografie; detaillierte bibliografische Daten sind im Internet über http://dnb.d-
nb.de/ abrufbar.

Imprint:

Copyright © 2011 GRIN Verlag GmbH
Druck und Bindung: Books on Demand GmbH, Norderstedt Germany
ISBN: 978-3-640-98963-8

This book at GRIN:

http://www.grin.com/en/e-book/177288/a-study-on-aspergillus-flavus

A Study on *Aspergillus flavus*

(Biochemical characterization of *Aspergillus flavus* isolates)

A Minor Project Report Submitted

By

Lisa Nathalie Tantio

Submitted in partial fulfillment of Degree of Master of Science to

University of Mysore

Department of Studies in Applied Botany & Biotechnology

University of Mysore, Manasagangothri

Mysore 570006

June 2011

CERTIFICATE

This is to certify that the minor project work report entitled "**A Study on** *Aspergillus flavus* (**Biochemical characterization of** *Aspergillus flavus* **isolates**)" submitted for the completion of second semester term to University of Mysore by **Ms. Lisa Nathalie Tantio** is the result of bonafide work carried out by her, in **Department of Studies in Applied Botany & Biotechnology, University of Mysore, Mysore** under my guidance during March-June 2011.

Date :

Place : Mysore

DECLARATION

I hereby declare that this project investigation report entitled "**A Study on** *Aspergillus flavus* (**Biochemical characterization of** *Aspergillus flavus* **isolates**)" submitted for the completion of second semester term to University of Mysore by **Ms. Lisa Nathalie Tantio** is the result of work carried done by me in **Department of Studies in Applied Botany & Biotechnology, University of Mysore, Mysore** under the guidance of Prof. S.R. Niranjana during March-June 2011.

Date : June 2011

Place : Mysore   **(Lisa Nathalie Tantio)**

ACKNOWLEDGEMENT

It is with a great pleasure that, I express my deep sense of respectful gratitude and sincere thanks to my guide **Prof. S.R. Niranjana**, professor, Department of Studies in Applied Botany and Biotechnology, University of Mysore, Manasagangothri, Mysore for his guidance and thoughts throughout my work.

I am grateful to our Chairman, **Prof. Rasheed Ahmad**, Department of Studies in Applied Botany and Biotechnology, University of Mysore, Manasagangothri, Mysore for the opportunity to complete this work.

I would like to extent my special thanks to my guide, **Ms. H.M. Navya** for her valuable suggestion and tremendous support.

I express my humble reverence to Mr. Naveen, Mr. Aiyaz, Ms. Rohini, and all research scholars for their constant guidance and help in shipping my work.

Last but not least, I would like to thanks each and every one who help me in finishing my project work.

TABLE OF CONTENTS

INTRODUCTION

Aspergillus flavus is the most widely known species of the genus *Aspergillus* which is known as a species in 1809 and first reported as a plant pathogen in 1920 (Leslie, 2008). Like other *Aspergillus* species, this fungus has a worldwide distribution due to its numerous conidia production, which easily disperses by air movements and possibly by insects. *Aspergillus flavus* is mainly a saprophyte in the soil, where it plays a major role as a nutrient recycler, supported by plant and animal debris and contaminates a wide variety of agricultural products in the field, storage areas, processing plants, and during distribution. The ability of *A. flavus* to survive in unfavorable conditions allows it to easily out-compete other organisms for substrates in the soil or plant. It grows better with water activity (a_w) ranges from 0.86 – 0.96 with an optimum temperature of 37 °C, but able to grow in the range of 12 – 48 °C. This optimum temperature contributes its pathogenicity in human being. (Hedayati *et al.*, 2007; Ruiqian *et al.*, 2004)

Aspergillus flavus belongs to Subgenus *Circumdati* sections *Flavi* and the taxonomy of this species is as follows:

Domain: Eukaryote

Kingdom: Fungi

Phylum: Ascomycota

Class: Eurotiomycetes

Order: Eurotiales

Family: Trichocomaceaae

Genus: Aspergillus

Species: *Aspergillus flavus* Link.

In general, *A. flavus* colony appear as a velvety, yellow to green or brown mold with colorless or sandy beige reverse. Old colony appears as dark green. The shape is smooth and some have radial wrinkles. Conidiophores are heavy walled, uncolored, coarsely roughened, usually less than 1 mm (400-800 μm) in length and are often rough just beneath the globose vesicles. Vesicles are elongate when young, later becoming subglobose or globose, varying from 10 to 65 mm in diameter. Phialides are uniseriate (single layered) or biseriate (two layered). The primary branches are up to 10 mm in length, and the secondary up to 5 mm in length (Hedayati *et al.*, 2007 & Ruiqian *et al.*, 2004).

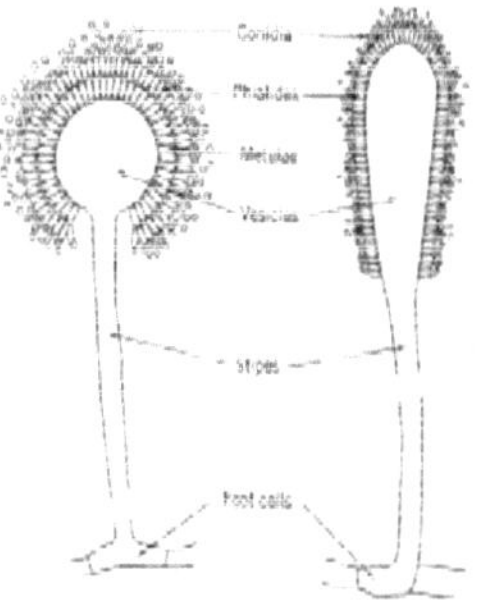

Fig: 1. Characteristics of *Aspergillus* conidiophores (Klich, 2007)

Many fungal cultures produce toxic secondary metabolite called mycotoxin. They are broadly classified as aflatoxins, trichothecenes, fumonisins, zearalenone, ochratoxin A and ergot alkaloids, which pose the greatest potential risk to plant, human, and animal health. Some mycotoxins that are produced by *A. flavus* include β-nitropropionic acid, aflatoxin B_1, aflatoxin B_2, aflatoxin B_{2a}, aflatoxin G_1, aflatoxin G_2, aflatoxin M, aflatrem, aspertoxin, cyclopiazonic acid, gliotoxin, sterigmatocystin, and versicolorin A.

Another species such as *A. parasiticus*, *A. nomius*, *A. pseudotamarii*, *A. bombycis*, and *A. toxicarius* also produces aflatoxin, difuranocoumarin derivative produced by a polyketide pathway. Four major types of aflatoxins have been characterized, B_1, B_2, G_1, and G_2. Aflatoxin B and aflatoxin G resemble the blue and green fluorescent observed under ultraviolet light, while the subscript designates relative chromatographic mobility. Aflatoxin M is mainly found in dairy product, and hence the M initial is taken from milk. Apart from the above types, aflatoxin B_1 is the most pathogenic compound. (Hedayati *et al.*, 2007; Klich, 2007).

Most *A. flavus* produces aflatoxin B_1 and B_2. Aflatoxin is harmful in plant, human and animal health. Some commodities that have been found contaminated are peanuts, cotton, corn, cereals, dried fruits, oilseed, wheat, rice, cottonseed, copra, nuts, coffee bean, dry bean, soybean, sorghum, barley, various foods, milk, eggs, cheese, and figs. In corn, *A. flavus* causes an ear rot.

In peanuts, it causes a rot in mature peanuts and a seedling disease known as yellow mould of seedlings or aflaroot. The symptoms include necrotic lesions, chlorosis above - ground parts and lack of development of secondary roots (aflaroot). In cotton, *A. flavus* affects cotton quality by causing boll rot. Furthermore, aflatoxin B_1 has been shown to inhibit seed germination of some other crop seeds, including wheat, corn, mustard, mung and gram (Klich, 2007).

In animals, birds, fish, and mammals (young pigs, pregnant sows, dog, calf, mature cattle, sheep, cat, monkey) have been reported infected by aflatoxin. The pathological effects are hepatotoxicity (liver damage), bile duct hyperplasia, hemorrhage (intestinal tract and kidneys), and liver tumors. In human, aflatoxin causes chronic cavitary pulmonary aspergillosis (CCPA) and aspergilloma (clump of fungus in body cavity, i.e. lung), allergic bronchopulmonary aspergillosis (ABPA) and allergens, keratitis and endophthalmitis, cutaneous infection, wound infection, endocarditis and pericarditis, central nervous system (CNS) infection, rhinosinusitis, allegic fungal sinusitis (AFS) and sinus aspergilloma, osteoarticular infection, and urinary tract infection. (Hedayati *et al.*, 2007).

Characterization study of *Aspergillus flavus* on its morphological, physiological, and biochemical aspect is important to identify this pathogenic fungi for the management of various human, animal, and plant diseases as well for surveillance, and other epidemiological study (Kiba *et al.,* 2007).

Aside from the above studies, the study in enzymes produce by this fungus is important because they play a major role in host-pathogen relationship. Numerous cell wall degrading enzymes can be secreted by pathogens to breach and use the plant cell wall as nutrient sources that eventually lead to develop inedible, undesirable quality, and soft rot spoilage. Pectinases are the first enzymes secreted by fungal pathogens when they attack plant cell walls. These enzymes weaken the plant cell wall and expose other polymers to degradation by hemicelluloses, cellulose, amylase, lipase, and protease (Hindi *et al.*, 2011). Fungi also produce catalase that reacts with and neutralize hydrogen peroxide produced by plant as one of the first defense response against infecting pathogens (Agrios, 2005).

In this study, 20 *A. flavus* isolates were first differentiated based on their toxic nature then they were characterized based on their morphological physiological and biochemical parameters to know is there any variation among the toxic and nontoxic isolates with respect to above said parameters.

MATERIALS AND METHODS

Aspergillus flavus **isolates**.

Twenty *A. flavus* isolates were collected from culture collection of Department of studies in Biotechnology, University of Mysore, Mysore and they were originally isolated from groundnut seeds that have been collected from different agroclimatic regions of India. The collected isolates were maintained on PDA (potato dextrose agar) slants at 4 °C for further use.

Differentiation of the isolates (aflatoxin study).

All the isolates were differentiated into toxigenic and non toxigenic by their ability to produce aflatoxins using HPTLC analysis. For HPTLC analysis, *A. flavus* isolates were grown on PDB (50 ml) for 4 days at 28±2° C. After incubation period, broth was separated from mycelia by filtration. Aflatoxins in the broth were extracted twice with equal volume of chloroform and evaporated to dryness. The residue was redissolved in 50 µl chloroform and applied onto HPTLC plate. Twenty µl of this chloroform extract was spotted on the base line of HPTLC plate (20 x 20 cm). Spotted plates were developed in Toluene: Ethyl acetate: Formic acid (10:8:2) solvent system for approximately 20 min so that the solvent front moves about 15 cm. Plates were dried and observed under long wavelength UV (365 nm) light fitted in a black cabinet (Singh *et al.,* 1999). Plates were observed for blue and green fluorescence at Rf-value 0.5 and 0.45 respectively. HPTLC analysis showed the presence of aflatoxin B_1, B_2, G_1 and G_2 in aflatoxigenic strains.

Morphological characterization of *A. flavus* **isolates.**

Four different culture media, PDA, CZ, YESA and AFPA were used for the morphological characterization of *A. flavus* isolates. Each medium was sterilized at 121 °C for 20 min and poured into pre sterilized petri dishes. Each of the isolate was inoculated on one plate of each media at the centre of the petri dish. After inoculation, all plates were incubated for seven days at 25±2 °C and observed for the macro morphological characters like colony color, colony reverse, diameter, growth, texture and margin of the colony (Batista *et al.,* 2008).

Physiological characterization of *A. flavus* isolates.

Physiological characterization for all the isolates was done by growing them on PDA medium with different incubation temperatures (4, 15, 30 and 40 °C) for 7 days and observed for the growth and sporulation of the isolate.

Biochemical characterization of *A. flavus* isolates.

For biochemical characterization, *A. flavus* isolates were inoculated into five different solid media for amylase, protease, lipase, pectinase and cellulase activity. Plates were kept at 25 ± 2 °C for 3-5 days and growth was observed.

Amylolytic activity. The ability to degrade starch was used as the criterion for determination of ability to produce amylases. The medium used contained $NaNO_3$, 1 g ; K_2HPO_4, 1 g ; $MgSO_4.7H_2O$, 1 g ; $FeSO_4$, 0.04 g ; soluble starch 20 g ; agar 25 g, and distilled water 1 litre. After 3-5 days of incubation, 5-8 pieces of iodine crystals were dropped onto the plates lid. A clear zone around the colony in an otherwise blue medium indicated amylolytic activity. The diameter of clear zone and diameter of fungal growth were measured.

Proteolytic activity. The medium used contained $NaNO_3$, 2 g ; K_2HPO_4, 1 g ; $MgSO_4.7H_2O$, 0.5 g ; KCl, 0.5 g ; $FeSO_4.7H_2O$, 0.01 g ; sucrose 30 g ; skim milk powder 10 g ; agar 20 g, and distilled water 1 litre. After incubation, complete degradation of protein in skim milk was observed as a clearing in the somewhat opaque agar around colonies. The diameter of clear zone and diameter of fungal growth were measured.

Lipolytic activity. The medium used contained peptone, 10 g ; NaCl, 5 g ; $CaCl_2.2H_2O$, 0.1 g ; agar 20 g, Tween 20, 10 ml, and distilled water 1 litre. The formation of lipolytic enymes was seen as either a visible precipitate due to the formation of the calcium salt of the lauric acid liberated by the enzyme, or as a clearing such as precipitate around a colony due to complete degradation of the salt of the fatty acid. Sorbitan monolaurate (Tween 20) was used as a source of fatty acid.

Pectolytic activity. The medium used contained yeast extract, 2 g ; pectin, 10 g ; K_2HPO_4, 1.5 g ; $MgSO_4.7H_2O$, 1.5 g ; agar, 25 g, and distilled water 1 litre. Following incubation, plates were flooded with a 1% aqueous solution of hexadecyltrimethylammonium bromide (cetrimide). This reagent precipitates intact pectin in the medium and, thus, clear zones around a colony in an otherwise opaque medium indicated degradation of the pectin.

Cellulolytic activity. The medium used contained CMC (carboxymethylcellulase), 10 g ; NaNO$_3$, 6.5 g ; K$_2$HPO$_4$, 6.5 g ; yeast extract, 0.3 g ; KCl, 6.5 g ; MgSO$_4$.7H$_2$O, 0.3 g ; glucose, 0.65 g ; agar 17.5 g ; distilled water 1 litre. After incubation, plates were flooded with a 1% congo red solution for 10-15 minutes, then de-stained with 1 M NaCl for 30 minutes until a clear zone can be seen. Unstained areas indicate where the CMC has been broken down to β 1,4 glucans that contain seven or fewer glucose residues. The diameter of the clear zone can be measured to provide a quantitative comparison of cellulolytic activity.

Catalytic activity. For catalase activity, the isolates were grown in potato dextrose broth (PDB) for 10 days. After incubation period, the mycelial mats were removed, and stored at -20 ℃. The frozen mycelial mat was ground with pre-chilled pestle and mortar using phosphate buffer (0.1 M; pH 6.0) and further lyses of cell wall was done by ultra sonication. The lysed cell suspension was centrifuged at 8,000 rpm for 10 min at 4 ℃ and the supernatant obtained was used as enzyme source. Catalase activity was determined spectrophotometrically to measure the decrease in absorbance at 240 nm during decomposition of H$_2$O$_2$ by the enzyme. One unit of catalase decomposed 1 μM of H$_2$O$_2$ in 1 min at 25 ℃.

The catalase isoform profiles of fungal cell lysates were examined by discontinuous native polyacrylamide gel electrophoresis (Native-PAGE). Fungal cell lysates (60 μg protein) were loaded onto 8.0 % (w/v) polyacrylamide gel with a vertical mini-gel electrophoresis unit. The electrode buffer was Tris-base (3.0 g Tris-base, 7.2 g glycine and 1 l distilled water, pH 8.3). Electrophoresis was performed at a constant voltage of 50 V initially for 1 h, and subsequently 100 V to complete electrophoresis. After electrophoresis, the gel was removed carefully and washed with distilled water to remove any bound salt. The gel was then treated with 0.003 % H$_2$O$_2$ for 10 min followed by washing in distilled water. Finally, it was stained with 1% potassium ferricyanide and 1% ferric chloride. The enzyme appeared as yellow bands on a dark green background. The reaction was halted by adding tap water, and the gel was then photographed.

RESULTS

Differentiation of aflatoxgenic and nonaflatoxigenic isolates by HPTLC

Presence of aflatoxin on HPTLC plate after development was visualized under long wavelength UV (365 nm) light. Among the 20 of isolates tested, 12 isolates of *Aspergillus flavus* were toxigenic by fluorescing blue spots for the presence of AFB1/AFB2 or both and 8 isolates were found to be non toxigenic without any fluorescence spot in comparison with the standard **(Table: 1 Fig: 2)**.

Morphological and physiological characterization of *Aspergillus flavus* isolates.

Aspergillus flavus isolates were characterized morphologically with four different culture media PDA, CZ, YESA and AFPA **(Table: 2 and 3; Fig: 3 and 4).** CZ media was recorded to be with faster growth rate and sporulation compare to others. Whereas, slow growth rate and much decreased rate of sporulation was found in AFPA media. With all the studied isolates, PDA grown isolates showed dark green and light green in the obverse and reverse side. Yellowish green to light green was the color of the isolates on CZ media. Light yellow color was appeared in all the YESA grown isolates. Almost all the isolates, which were grown on AFPA media, showed light pink of the mycelium with very less sporulation and orange reverse. The growth and sporulation of *A. flavus* was maximum at 30 °C.

Biochemical characterization of *A. flavus* isolates.

Amylase activity was shown by only 12 isolates and there is no clear zone in the remaining 8 isolates. Five isolates did not produce any protease among the 20 tested. Where as 18 isolates were positive for pectin methyl production.

In case of lipase, the ability to form white precipitate by isolates was estimated in terms positive and negative. Isolates AF10, AF11, AF19 did not show any white precipitation and all other isolates showed white precipitate due to the activity of lipase.

Cellulose activity was very low in all the tested isolates and only 2 (AF7 and AF14) isolates exhibited the activity of cellulose in culture plates **(Table: 4)**.

Specific activity of catalase was calculated for all the isolates and all the isolates showed varied level of activity. The highest activity (210 mg/ml) was recorded by AF3 isolate and the least activity (1.58 mg/ml) in AF12 **(Fig: 5 and 6; Table: 5)**.

Protein bands showing catalase activity were observed. Normally one isoform was obtained in all the isolates. Protein estimation for all enzyme assay was carried out following the standard procedure of Bradford (1976) using BSA (Sigma, USA) as a standard.

Table: 1. Detection of aflatoxins in *Aspergillus flavus* isolates by high performance thin layer chromatography

Sl. No.	Name of the isolate	HPTLC
1	AF1	AFB_1
2	AF2	AFB_1, AFB_2
3	AF3	AFB_1, AFB_2
4	AF4	AFB_1, AFB_2
5	AF5	AFB_1, AFB_2
6	AF6	AFB_1, AFB_2
7	AF7	AFB_1, AFB_2
8	AF8	AFB_1, AFB_2
9	AF9	AFB_1, AFB_2
10	AF10	AFB_1, AFB_2
11	AF11	-
12	AF12	-
13	AF13	-
14	AF14	AFB_1
15	AF15	-
16	AF16	-
17	AF17	-
18	AF18	-
19	AF19	AFB_1, AFB_2
20	AF20	-

AFB_1: aflatoxin B_1; AFB_2: aflatoxin B_2; -: negative for aflatoxin production

Table: 2. Morphological characteristics of *Aspergillus flavus* isolates

Name of the isolate	Media	Color of the colony		Colony diameter (cm)	Colony Texture	Colony Margin	Sporulation
		Obverse	Reverse				
AF7	PDA	Green with white edge	Creamish white	4.6	Powdery	Wavy	Less
	YESA	White	Yellow	6.8	Cotton	Entire	More
	CZ	Yellow with white edge	Cream	6.5	Powdery	Entire	More
	AFPA	White with transparent edge	Red	6.0	Cotton	Entire	More
AF9	PDA	Green with white edge	Creamish white	4.7	Powdery	Wavy	Less
	YESA	White cotton with yellow centre	Yellow	6.5	Cotton	Wavy	More
	CZ	Yellow	Cream	6.5	Powdery	Wavy	More
	AFPA	White with cream edge	Red	6.2	Cotton	Wavy	More
AF12	PDA	Green with white edge	Creamish white	4.6	Powdery	Wavy	Less
	YESA	White cotton with yellow centre and green	Yellow	6.5	Powdery	Wavy	More
	CZ	Yellow with white edge	Yellow with white edge	6.5	Powdery	Wavy	More
	AFPA	Pinkish white edge with brown centre	Red	6.2	Cotton	Entire	More

Table: 2. Morphological characteristics of *Aspergillus flavus* isolates *(continued)*

Name of the isolate	Media	Color of the colony		Colony diameter (cm)	Colony Texture	Colony Margin	Sporulation
		Obverse	Reverse				
AF13	PDA	Green with white edge	Creamish white	4.6	Powdery	Wavy	Less
	YESA	White	Yellow	6.1	Cotton	Entire	More
	CZ	Yellow with white edge	Cream	7.0	Powdery	Entire	More
	AFPA	Pinkish white edge with brown centre	Red	6.1	Cotton	Entire	More
AF15	PDA	Green with white edge	Creamish white	4.6	Powdery	Wavy	Less
	YESA	White cotton with yellow centre	Yellow	7.6	Cotton	Wavy	More
	CZ	Yellow with white edge	Cream	6.6	Powdery	Entire	More
	AFPA	Pinkish white edge with brown centre	Red	7.1	Cotton	Wavy	More
AF16	PDA	Green with white edge	Yellow	4.4	Powdery and cotton	Wavy	More
	YESA	White cotton with yellow centre	Yellow	7.0	Powdery	Wavy	More
	CZ	Yellow	Yellow	6.5	Powdery	Entire	More
	AFPA	Pinkish white edge with brown centre	Red	5.5	Cotton	Entire	More

Table: 2. Morphological characteristics of *Aspergillus flavus* isolates *(continued)*

Name of the isolate	Media	Color of the colony		Colony diameter (cm)	Colony Texture	Colony Margin	Sporulation
		Obverse	Reverse				
AF17	PDA	Green with white edge	Creamish white	4.8	Powdery	Entire	Less
	YESA	White cotton with yellow centre	Yellow	8.8	Cotton	Entire	More
	CZ	Yellow with white edge	Yellow	7.1	Powdery	Entire	More
	AFPA	Pinkish white edge with brown centre	Red	6.0	Cotton	Entire	More
AF18	PDA	Green with white edge	Creamish white	4.5	Powdery	Wavy	Less
	YESA	White	Yellow	6.3	Cotton	Wavy	More
	CZ	Yellow with white edge	Cream	7.3	Powdery	Wavy	More
	AFPA	Pinkish white edge with brown centre	Red	6.1	Cotton	Entire	More
AF19	PDA	Green with white edge	Creamish white	4.3	Powdery	Wavy	Less
	YESA	White	Yellow	5.1	Cotton	Wavy	More
	CZ	Yellow with white edge	Cream	6.5	Powdery	Wavy	More
	AFPA	-	-	-	-	-	-
AF20	PDA	Green with white edge	Creamish white	5.5	Powdery	Wavy	Less
	YESA	White	Yellow	6.8	Cotton	Entire	More
	CZ	Yellow with white edge	Cream	6.8	Powdery	Entire	More
	AFPA	Pinkish white edge with brown centre	Red	6.9	Cotton	Entire	More

PDA: Potato Dextrose Agar; YESA: Yeast Extract Salt Agar; CZ: Czapek Dox; AFPA: *Aspergillus flavus parasiticus* Agar

Table: 3. Physiological characteristics of *Aspergillus flavus* isolates

Name of the isolate	Temperature (°C)	Color of the colony		Diameter of the colony (cm)
		Obverse	Reverse	
AF7	4	NG	NG	-
	15	Green with white edge	Cream	5.3
	30	Green with white edge	Creamish white	4.5
	40	Opaque	Brown	1.3
AF9	4	NG	NG	-
	15	Green with white edge	Cream	5.0
	30	Green with white edge	Creamish white	4.8
	40	Opaque	Brown	0.7
AF12	4	NG	NG	-
	15	Green with white edge	Cream	4.9
	30	Green with white edge	Creamish white	4.4
	40	Opaque	Brown	1.4
AF13	4	NG	NG	-
	15	Green with white edge	Cream	5.0
	30	Green with white edge	Creamish white	4.6
	40	Opaque	Brown	1.1
AF15	4	NG	NG	-
	15	Green with white edge	Cream	4.9
	30	Green with white edge	Creamish white	4.6
	40	Opaque	Cream	0.7
AF16	4	NG	NG	-
	15	Green with white edge	Cream	4.2
	30	Green with white edge	Creamish white	4.3
	40	Opaque	Brown	1.3
AF17	4	NG	NG	-
	15	Green with white edge	Cream	5.1
	30	Green with white edge	Creamish white	5.5
	40	Opaque	Cream	2
AF18	4	NG	NG	-
	15	Green with white edge	Cream	5.1
	30	Green with white edge	Creamish white	4.6
	40	Opaque	Brown	1.3
AF19	4	NG	NG	-
	15	Green with white edge	Cream	5.1
	30	Green with white edge	Creamish white	4.6
	40	Opaque	Brown	0.8
AF20	4	NG	NG	-
	15	Green with white edge	Cream	5.0
	30	Green with white edge	Creamish white	4.7
	40	Opaque	Cream	0.5

NG = no growth

Table: 4. Extracellular enzymes activity of *Aspergillus flavus* isolates

Sl. No.	Name of the isolate	Enzyme activity (DCZ/DFC)				Lipase
		Amylase	PME	Protease	Cellulase	
1	AF1	1.07	1.1	1.04	-	+
2	AF2	1.07	1.1	1.03	-	+
3	AF3	1.12	1.1	1.1	-	+
4	AF4	1	1.1	-	-	+
5	AF5	1	1.1	1.08	-	+
6	AF6	1	1.1	1.06	-	+
7	AF7	1.2	1.08	-	1.1	+
8	AF8	1	1.09	-	-	+
9	AF9	1.2	1.1	1.08	-	+
10	AF10	1	1.1	1.03	-	-
11	AF11	1.1	1.08	1.08	-	-
12	AF12	1.1	1.1	1.08	-	+
13	AF13	1.1	1	1.08	-	+
14	AF14	1	1.1	-	1.1	+
15	AF15	1.03	1.1	1.1	-	+
16	AF16	1	1	1.06	-	+
17	AF17	1	1.1	1.06	-	+
18	AF18	1.1	1.1	1.06	-	+
19	AF19	1.1	1.1	1.05	1.03	-
20	AF20	1.1	1.1	-	1.05	+

PME: pectin methyl esterase; DCZ: diameter of clear zone; DFC: diameter of fungal colony

Table: 5. Specific activity catalase of *Aspergillus flavus* isolates

Sl. No	Name of the isolate	Specific activity of catalase (mg/ml)
1	AF1	22.97
2	AF2	15.8
3	AF3	210
4	AF4	20.3
5	AF5	8.92
6	AF6	3.17
7	AF7	22.0
8	AF8	73.17
9	AF9	3.33
10	AF10	60.6
11	AF11	31.0
12	AF12	1.58
13	AF13	7.23
14	AF14	7.82
15	AF15	12.5
16	AF16	5.46
17	AF17	10.70
18	AF18	109.4
19	AF19	3.50
20	AF20	6.8

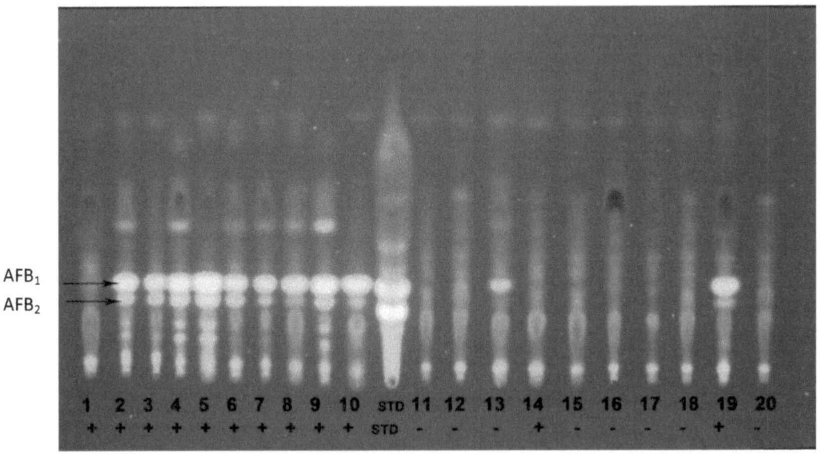

Fig: 2. Detection of aflatoxigenic *Aspergillus flavus* isolates from groundnut seed samples by HPTLC method.

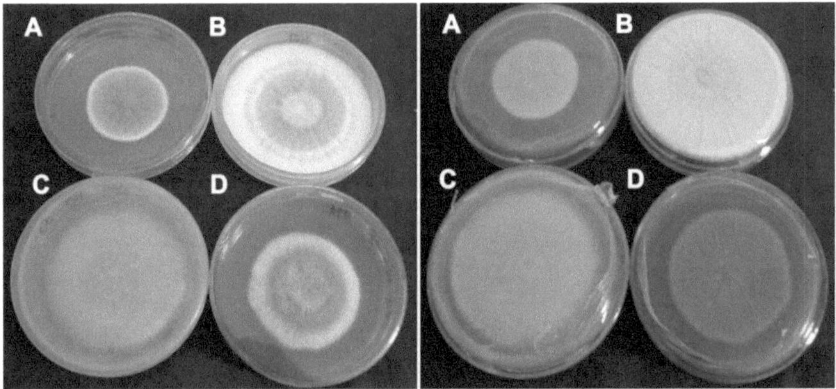

Fig: 3. Morphological Characterization of *Aspergillus flavus* on A: PDA, B: YESA, C: CZ and D: AFPA Media.

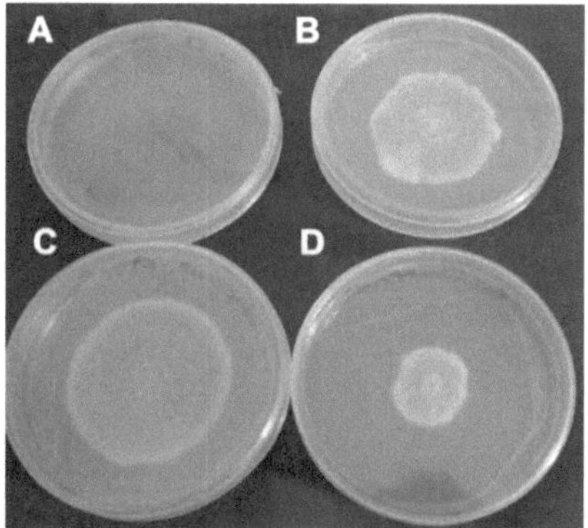

Fig: 4. Physiological features of *A, flavus* isolates at different temperature range -
A: 4 ^{o}C, B: 15 ^{o}C, C: 30 ^{o}C, and D: 40 ^{o}C

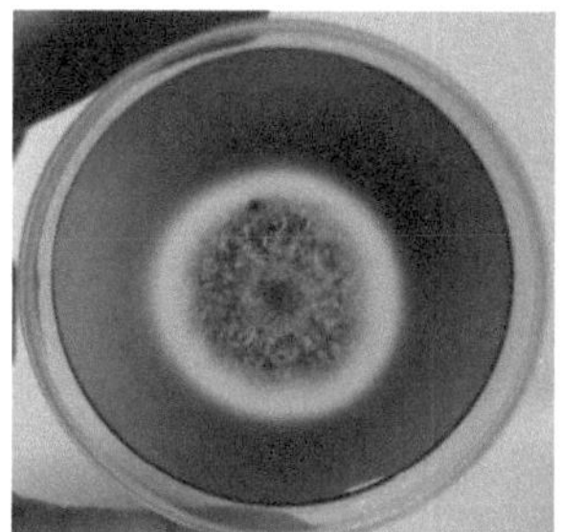

Amylase activity shown by *A. flavus* isolate in culture plate

Protease activity shown by *A. flavus* Isolate in culture plate

Fig: 5. Extracellular enzyme activity by *A. flavus* isolates

Lipase activity shown by
***A. flavus* isolate in culture plate**

**Pectin methyl esterase activity shown
by *A. flavus* isolate in culture plate**

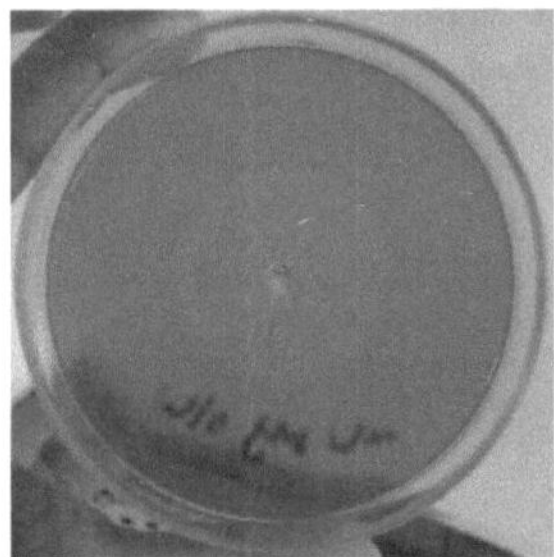

**Cellulase activity shown by *A.
flavus* isolate in culture plate**

Fig: 5. Extracellular enzyme activity by *A. flavus* isolates *(continued)*

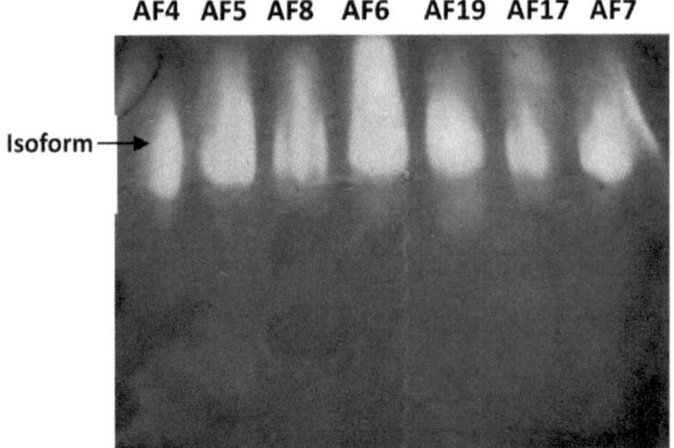

**Fig: 6. Native page gel showing isoform obtained by *Aspergillus flavus* isolates by the
activity of catalase**

DISCUSSIONS

The presence of aflatoxin, a secondary metabolite produced by *A. flavus* was determined using TLC assay and observed under UV light (Figure 2). Green fluorescent was not observed in all isolates, thus indicating that this fungus does not produce aflatoxin G. This trait is useful to distinguish between *A. flavus* and *A. parasiticus*, in which the latter species produce both aflatoxin B and aflatoxin G. Instead of showing green fluorescent, 70% of the isolates showed blue fluorescent and from these isolates, 2 isolates (AF1 and AF14) showed only aflatoxin B_1 production, the remaining isolates showed the production of aflatoxin B_1 and B_2. Thirty percent of (AF11, AF12, AF13, AF15, AF16, AF17, AF18 and AF20) of the total isolates did not produce aflatoxin. The retention factor for each spots from the point of spotting (in cm) was calculated using the following formula and compared with aflatoxin standard:

$$Rf = \frac{\text{distance travelled by the solute}}{\text{distance travelled by th solvent}}$$

The presence of aflatoxin B_1 and B_2 were distinguishable from the Rf value, in which the latter has lower Rf value (0.24) compare to aflatoxin B_1 (0.35).

Morphological, physiological, biochemical characterization, and aflatoxin studies have been conducted towards twenty *Aspergillus flavus* isolates collected from groundnut seed all across India. Morphological study was performed to identify *A. flavus* by using four different, either general media for fungi (PDA, CZ, YESA) or selective media for *A. flavus* (AFPA). Colonies that appear on each media are specific based on the nutrient in the media.

The morphological features of *A. flavus* isolates are represented in Table 2 and Figure 3. AFPA medium, a differential and selective medium for *A. flavus* is use to distinguish this fungus with *A. parasiticus*. *Aspergillus flavus* isolate will turn red color in reverse. Potato dextrose, potato flake, malt extract, inhibitory mould agar, or similar sporulation agars as primary isolation media for *Aspergillus* spp. may speed growth rate and the production of conidia. The addition of antibacterial agents to isolation media helps reduce time to identification by inhibiting bacterial overgrowth and reducing the need for subculture (Diba, 2007). Czapek dox agar (CZA) medium is a semi-synthetic medium used for fungal cultivation with sucrose as the carbon source and sodium nitrate as the sole nitrogen source. Dipotasium phosphate buffers the medium. Magnesium sulphate and ferrous sulphate as cation source, while potassium chloride acts as an essential ion. This media is useful to differentiate *A. flavus* and *A. parasiticus* colony which appear as yellow-green colony and darker-green colony, respectively.

The purpose of physiological study is to determine the optimal temperature for fungal growth. In this study (Table 3), fungal growth was observed under different temperature, 4 $^{\circ}$C, 15 $^{\circ}$C, 30 $^{\circ}$C, and 40 $^{\circ}$C. *Aspergillus flavus* can grow in temperature range from 15 – 40 $^{\circ}$C, but the optimal growth is observed in 30 $^{\circ}$C. No growth has been observed at 4 $^{\circ}$C for all plates. In addition, the diameter of the colony (Figure 4) is larger in 30 $^{\circ}$C compare to 40 $^{\circ}$C, suggesting that this fungus grows better at 30 $^{\circ}$C environment.

Fungi are highly successful in survival because of their physiological versatility. They thrive well in unfavorable environments due to their efficient enzyme system, mainly the extracellular enzymes. Aside from their function as a virulence factor in plant infection, these enzymes are useful for biotechnological purposes, such as amylase for bread making and production of glucose, protease for pharmaceutical industry and detergent industry, pectinase for fruit juice processing, lipase and cellulose for food industry application. Thereby, the study of extracellular enzyme is to be the basis in fungal studies.

Amylase, protease, pectinase, lipase, and cellulose production were detected using agar media containing the respective substrate. For amylolytic, proteolytic, pectinolytic, and cellulolytic study, positive result was considered if the clear zone diameter is larger than the fungal growth zone (Gopinath *et al.*, 2005). Positive result in lipolytic assay indicates the formation of calcium salt crystal.

As per table 4, eight isolates (AF4, AF5, AF6, AF8, AF10, AF14, AF16 and AF17) showed no activity of amylase from total 20 isolates. Pectinolytic activity was observed in all isolates, except AF13 and AF16. 1.5% (3 isolates, AF10, AF11 and AF19) showed negative lipolytic activity, and 25% (5 isolates, AF4, AF7, AF8, AF14, and AF20) showed no proteolytic activity. On the contrary, only 20% (AF7, AF14, AF19 and AF20) of the isolates showed cellulases production. The activity of the fungus to degrade cellulose is observed using congo red solution as an indicator. The specific colony in different media is represented in Figure 5.

As per table 5, the specific activity of catalase is determined using spectrophotometriy at 240 nm. High amount of specific activity represents high activity to degrade cellulose presents in plant cell wall. The highest catalase activity achieved by AF3 isolate, 210 mg/ml, whilst the lowest activity obtained by AF13, 1.58 mg/ml. Furthermore, the formation of isoform of the enzyme is determined using NATIVE-PAGE. In Figure 6, the presence of isoenzyme is detected as two bands. Isoenzyme have same function but in different conformation.

In summary, morphological, physiological, and biochemical studies are important to identify plant pathogenic fungus such *Aspergillus flavus*. Aflatoxin, being the widely produced mycotoxin of *A. flavus* is easily contaminates food and soil, thus its detection is required. Furthermore, by screening the extracellular enzymes activity, one can understand the ability of particular fungus to produce certain enzymes that have major contribution in biotechnology.

REFERENCE

Agrios, G. N. 2005. Plant pathology (5th ed, p. 229). Amsterdam, Netherland: Elsevier Academic Press.

Diba, K., Kordbacheh, P., Mirhendi, S. H., Rezaie, S., & Mahmoudi, M. 2007. Identification of *Aspergillus* species using morphological characteristics. *Pakistan Journal of Medical Science*. 23(6): 867-872.

Hedayati, M. T., Pasqualotto, A. C., Warn, P. A., Bowyer, P., and Denning, D. W. 2007. *Aspergillus flavus*: human pathogen, allergen, and mycotoxin producer. *Microbiology*. 153, 1677-1692.

Hesseltine, C. W., Shotwell, O. L., Ellis, J. J. and Stubblefield, R. D. 1966. Aflatoxin formation by *Aspergillus flavus*. *Bacteriological Reviews*. 30(4): 795-805.

Hindi, R. R., Al-Najada, A. R., and Mohamed, S. A. 2011. Isolation and identification of some fruit spoilage fungi: screening of plant cell wall degrading enzymes. African Journal of Microbiology Research. 5(4): 443-448.

Klich, M. A. 2007. *Aspergillus flavus*-the major producer of aflatoxin. Molecular Plant Pathology. 8(6): 713-722.

Riuqian, L., Qian, Y., Thanaboripat, D. and Thansukon, P. 2004. Biocontrol of *Aspergillus flavus* and aflatoxin production. *KMITL Science and Technology Journal*. 4(1).

http://books.google.com

www.himedialabs.com/TD/M075.pdf

http://en.wikipedia.org/wiki/Aspergillus_flavus

http://www.trilogylab.com/pdf/Mycotoxin_CAST_Report.pdf

http://www.aspergillus.org.uk/indexhome.htm?secure/metabolites/list_by_secmet.php~main